<u>Preface</u>

This book is solely created for educational purpose. It should be utilized for the same. Any method including copying and illegal selling will cause copyright infringement.

I hope you enjoyed the First Edition. I had already discussed Number System, Percentage, Divisibility rules in my first edition. We will continue with a little bit advanced application of Percentage. Now let's continue.

Topics Covered

Profit and Loss

Averages

Ratio and Proportion

Profit and Loss

Profit and Loss

To understand profit and loss we need to understand some other terms as well.

Buying Price/Cost Price- The Price at which goods/services are bought/manufactured.

Selling Price- The Price at which goods/Services are sold.

Marked Price- The Price printed on the product for selling.

- Profit (or Loss) = SP − CP (profit is made only when SP is greater than CP)
- Profit % (or Loss %) = (Actual Profit/Loss ÷ CP) × 100%
- CP = (SP × 100) ÷ (100 + profit %)
- SP = (100 + profit %) × C.P ÷ 100
- Actual Discount = MP − SP
- Discount % = (Actual Discount ÷ MP) × 100%
- SP = (MP − Discount %) of MP

Problem Set

Question

If the cost price of 20 articles is equal to the selling price of 16 articles, what is the percentage profit or loss made by the merchant?

1. 20% Profit
2. 25% Loss
3. 25% Profit
4. 33.33% Loss

Correct Answer

25% Profit. Choice (3)

Explanatory Answers

Let Cost price of 1 article be Re.1.

Therefore, Cost price of 20 articles = Rs.20.

Selling price of 16 articles = Rs.20

Therefore, Selling price of 20 articles = $\left(\dfrac{20}{16}\right)*20 = 25$

Profit = Selling price - Cost price

= 25 - 20 = 5

Percentage of profit = $\dfrac{\text{Profit}}{\text{Cost Price}}*100$.

= $\dfrac{5}{20}*100$ = 25% Profit

Correct Answer (3)

Question

If the price of petrol increases by 25%, by how much must a user cut down on his petrol consumption so that his expenditure on petrol remains unchanged?

    1.   25%

2. 16.67%
3. 20%
4. 33.33%

Correct Answer

20%. Choice (3)

Explanatory Answers

Let the price of petrol be Rs.100 per litre. Let the user use 1 litre of petrol. Therefore, his expense on petrol = 100 * 1 = Rs.100

Now, the price of petrol increases by 25%. Therefore, the new price of petrol = Rs.125.

As he has to maintain his expenditure on petrol constant, he will be spending only Rs.100 on petrol.

Let 'x' be the number of litres of petrol he will use at the new price.

Therefore, $125 * x = 100 \Rightarrow x = \dfrac{100}{125} = \dfrac{4}{5}$ =0.8 litres.

He has cut down his petrol consumption by 0.2 litres = $\dfrac{0.2}{1} * 100$ = 20% reduction.

There is a short cut for solving this problem.

If the price of petrol has increased by 25%, it has gone up $\dfrac{1}{4}$th of its earlier price.

Therefore, the % of reduction in petrol that will maintain the amount of money spent on petrol constant = $\dfrac{1}{4+1} = \dfrac{1}{5}$ 20%

i.e. Express the percentage as a fraction. Then add the numerator of the fraction to the denominator to obtain a new fraction. Convert it to percentage - that is the answer.

Correct Answer (3)
Question

A piece of equipment cost a certain factory Rs. 600,000. If it depreciates in value, 15% the first year, 13.5 % the next year, 12% the third year, and so on, what will be its value at the end of 10 years, all percentages applying to the original cost?

(1) 2,00,000
(2) 1,05,000
(3) 4,05,000
(4) 6,50,000

Explanatory Answer

Let the cost of an equipment is Rs. 100.
Now the percentages of depreciation at the end of 1st, 2nd, 3rd years are 15, 13.5, 12, which are in A.P., with a = 15 and d = - 1.5.

Hence, percentage of depreciation in the tenth year = a + (10-1) d
= 15 + 9 (-1.5) = 1.5
Also total value depreciated in 10 years = 15 + 13.5 + 12 + ... + 1.5 = 82.5

Hence, the value of equipment at the end of 10 years=100 - 82.5 = 17.5.
The total cost being Rs. 6,00,000/100 * 17.5 = Rs. 1,05,000.

Question
If a merchant offers a discount of 40% on the marked price of his goods and thus ends up selling at cost price, what was the % mark up?

1. 28.57%
2. 40%
3. 66.66%
4. 58.33%

Correct Answer - 66.66%. Choice (3).

Explanatory Answer

If the merchant offers a discount of 40% on the marked price, then the goods are sold at 60% of the marked price.

The question further states that when the discount offered is 40%, the merchant sells at cost price.

Therefore, selling @ 40% discount = 60% of marked price (M) = cost price (C)

ie., $\dfrac{60}{100} M = C$

or $M = \dfrac{100}{60} C$ or M = 1.6666C

i.e., a mark up 66.66%

Correct answer choice (3)

Question

If a merchant offers a discount of 30% on the list price, then she makes a loss of 16%. What % profit or % loss will she make if she sells at a discount of 10% of the list price?

1. 6% loss
2. 0.8% profit
3. 6.25% loss
4. 8% profit

Correct Answer - **8% profit**. Choice (4)

Explanatory Answer

Let the cost price of the article be Rs.100.
Let the List price of the article by "x".

Then, when the merchant offers a discount of 30%, the merchant will sell the article at x - 30% of x = 70% of x = 0.7x. ....(1)
Note: Discount is measured as a percentage of list price.

The loss made by the merchant when she offers a discount of 30% is 16%.
Therefore, the merchant would have got 100 - 16% of 100 = Rs.84 when she offered a discount of 30%. ....(2)
Note: Loss is always measured as a percentage of cost price.

Therefore, equating equations (1) and (2), we get
0.7x = 84
or x = 120.

If the list price is Rs.120 (our assumption of cost price is Rs.100), then when the merchant offers a discount of 10%, she will sell the article at
120 - 10%o of 120 = Rs.108.

As the cost price of the article was Rs.100 and the merchant gets Rs.108 while offering a discount of 10%, she will make a profit of 8%.

Correct answer choice (4)

Question
A merchant marks his goods up by 60% and then offers a discount on the marked price. If the final selling price after the discount results in the merchant making no profit or loss, what was the percentage discount offered by the merchant?
1. 60%
2. 40%
3. 37.5%
4. Depends on the cost price

Correct Answer - **37.5% discount**. Choice (3)
Explanatory Answer
Assume the cost price to be 100.
Therefore, the merchant's marked price will be 100 + 60% of 100
= 160

Now, the merchant offers a discount on the marked price. The discount results in the merchant selling the article at no profit or loss or at the cost price.

That is the merchant has sold the article at 100.

Therefore, the discount offered = 60.

Discount offered is usually measured as a percentage of the marked price.

Hence, % discount = $\dfrac{60}{160} * 100 = 37.5\%$

Correct answer choice (3)

Question

A merchant marks his goods up by 75% above his cost price. What is the maximum % discount that he can offer so that he ends up selling at no profit or loss?

1. 75%
2. 46.67%
3. 300%
4. 42.85%

Correct Answer - **42.85%**. Choice (4)

Explanatory Answer

Let us assume that the cost price of the article = Rs.100
Therefore, the merchant would have marked it to Rs.100 + 75% of
Rs.100 = 100 + 75 = 175.

Now, if he sells it at no profit or loss, he sells it at the cost price.
i.e. he offers a discount of Rs.75 on his selling price of Rs.175

Therefore, his % discount = $\dfrac{75}{175} * 100$ = 42.85%
Correct answer choice (4)

Question
A merchant marks his goods in such a way that the profit on sale
of 50 articles is equal to the selling price of 25 articles. What is his
profit margin?
    1.   25%
    2.   50%
    3.   100%
    4.   66.67%
Correct Answer - **100% profit**. Choice (3)
Explanatory Answer
Let the selling price per article be = Re. 1
Therefore, selling price of 50 articles = Rs.50

Profit on sale of 50 articles = selling price of 25 articles = Rs.25.
S.P = 50. Profit = 25. Therefore, CP = 50 - 25 = 25.

And % Profit = $\dfrac{Profit}{C.P} * 100 = \dfrac{25}{25} * 100 = 100\%$

Correct answer choice (3)

Question
Two merchants sell, each an article for Rs.1000. If Merchant A
computes his profit on cost price, while Merchant B computes his

profit on selling price, they end up making profits of 25%
respectively. By how much is the profit made by Merchant B
greater than that of Merchant A?

1. Rs.66.67
2. Rs.50
3. Rs.125
4. Rs.200

Correct Answer - **Rs.50**. Choice (2)

Explanatory Answer

Merchant B computes his profit as a percentage of selling price.
He makes a profit of 25% on selling price of Rs.1000. i.e. his
profit = 25% of 1000 = Rs.250

Merchant A computes his profit as a percentage of cost price.
Therefore, when he makes a profit of 25% or 1/4th of his cost
price, then his profit expressed as a percentage of selling price =

$\dfrac{1}{1+4} = \dfrac{1}{5}$ th or 20% of selling price.

So, Merchant A makes a profit of 20% of Rs.1000 = Rs.200.

Merchant B makes a profit of Rs.250 and Merchant A makes a
profit of Rs.200
Hence, Merchant B makes Rs.50 more profit than Merchant A.

Correct answer choice (2)

Question

A trader buys goods at a 19% discount on the label price. If he
wants to make a profit of 20% after allowing a discount of 10%,
by what % should his marked price be greater than the original
label price?

1. +8%
2. -3.8%
3. +33.33%
4. None of these

Correct Answer - **8% profit**. Choice (1)

Explanatory Answer

Let the label price be = Rs.100. The trader buys at a discount of 19%.
Hence, his cost = 100 - 19 = 81.

He wants to make a profit of 20%. Hence his selling price = 1.2 (81) = 97.2

However, he wants to get this Rs.97.2 after providing for a discount of 10%. i.e. he will be selling at 90% of his marked price.

Hence, his marked price M = $\dfrac{97.2}{0.9}$ = 108 which is 8% more than the original label price.

Correct answer choice (3)

Question

Rajiv sold an article for Rs.56 which cost him Rs.x. If he had gained x% on his outlay, what was his cost?

1. Rs.40
2. Rs.45
3. Rs.36
4. Rs.28

Correct Answer - **40**. Choice (1)

Explanatory Answer

x is the cost price of the article and x% is the profit margin.

Therefore, s.p = $x * \left(1 + \dfrac{x}{100}\right) = 56$

=> $x * \left(\dfrac{100 + x}{100}\right) = 56$

So, $100x + x^2 = 5600$.
Solving for 'x' , we get x = 40 or x = -140.
As the price cannot be a -ve quantity, x = 40.

The cost price is 40 and the markup is 40.

It is usually easier to solve such questions by going back from the answer choices as it saves a considerable amount of time.

Correct answer choice (1)

Question
A trader professes to sell his goods at a loss of 8% but weights 900 grams in place of a kg weight. Find his real loss or gain per cent.
1. 2% loss
2. 2.22% gain
3. 2% gain
4. None of these

Correct Answer - **2.22% gain**. Choice (2)

Explanatory Answer
The trader professes to sell his goods at a loss of 8%.
Therefore, Selling Price = (100 - 8)% of Cost Price
or SP = 0.92CP

But, when he uses weights that measure only 900 grams while he claims to measure 1 kg.
Hence, CP of 900gms = 0.90 * Original CP

So, he is selling goods worth 0.90CP at 0.92CP
Therefore, he makes a profit of 0.02 CP on his cost of 0.9 CP

$$\text{Profit \%} = \frac{SP - CP}{CP} \times 100$$

i.e., $\dfrac{0.92 - 0.90}{0.90} \times 100 = \dfrac{0.02}{0.90} \times 100 = 2\dfrac{2}{9}\%$ or 2.22%

Correct answer choice (2)

Question

A merchant buys two articles for Rs.600. He sells one of them at a profit of 22% and the other at a loss of 8% and makes no profit or loss in the end. What is the selling price of the article that he sold at a loss?

1. Rs.404.80
2. Rs.440
3. Rs.536.80
4. Rs.160

Correct Answer - **Rs.404.80** . Choice (1)

Explanatory Answer

Let C1 be the cost price of the first article and C2 be the cost price of the second article.
Let the first article be sold at a profit of 22%, while the second one be sold at a loss of 8%.

We know, C1 + C2 = 600.
The first article was sold at a profit of 22%. Therefore, the selling price of the first article = C1 + (22/100)C1 = 1.22C1
The second article was sold at a loss of 8%. Therefore, the selling price of the second article = C2 - (8/100)C2 = 0.92C2.

The total selling price of the first and second article = 1.22C1 + 0.92C2.

As the merchant did not make any profit or loss in the entire transaction, his combined selling price of article 1 and 2 is the

17

same as the cost price of article 1 and 2.

Therefore, $1.22C1 + 0.92C2 = C1 + C2 = 600$
As $C1 + C2 = 600$, $C2 = 600 - C1$. Substituting this in $1.22C1 + 0.92C2 = 600$, we get
$1.22C1 + 0.92(600 - C1) = 600$
or $1.22C1 - 0.92C1 = 600 - 0.92*600$
or $0.3C1 = 0.08*600 = 48$
or $C1 = 48/(0.3) = 160$.
If $C1 = 160$, then $C2 = 600 - 160 = 440$.
The item that is sold at loss is article 2. The selling price of article $2 = 0.92*C2 = 0.92*440 = 404.80$.

Note: When you actually solve this problem in CAT, you should be using the following steps only
$1.22C1 + 0.92C2 = C1 + C2 = 600$
$1.22C1 + 0.92(600 - C1) = 600$
$C1 = 48/(0.3) = 160$.
$C2 = 600 - 160 = 440$.
And the final step of the answer which is $0.92*440$ which you should not actually compute. As two of the answer choices (2) and (3) are either 440 or more, they cannot be the answers. The last one is way too low to be 92% of 440, therefore, the answer should be choice (1)

Question
A trader makes a profit equal to the selling price of 75 articles when he sold 100 of the articles. What % profit did he make in the transaction?
1. 33.33%
2. 75%
3. 300%
4. 150%
Correct Answer - **300% profit**. Choice (3)

Explanatory Answer

Let S be the selling price of 1 article.

Therefore, the selling price of 100 articles = 100 S. --(1)

The profit earned by selling these 100 articles = selling price of 75 articles = 75 S -- (2)

We know that Selling Price (S.P.) = Cost Price (C.P) + Profit -- (3)

Selling price of 100 articles = 100 S and Profit = 75 S from (1) and (2). Substituting this in eqn (3), we get

100 S = C.P + 75 S. Hence, C.P = 100 S - 75 S = 25 S.

Profit % = $\dfrac{Profit}{Cost.price} \times 100 = \dfrac{75S}{25S} \times 100 = 300\%$

Question

If a merchant makes a profit of 20% after giving a 20% discount, what should be his mark-up?

1. 20%
2. 40%
3. 50%
4. 60%
5. 48%

Correct Answer - **50%**. Choice (3)

Explanatory Answer

Assume cost price = 100.

As profit % = 20%, the profit made after the discount = 20% of 100 = Rs.20

So, selling price = Rs.120.

The merchant makes this profit of 20% after a discount of 20%.

Therefore, he is selling at Marked price - 20% of marked price = 80% of marked price.

i.e., 80% of marked price = Rs.120

$\dfrac{80}{100}$ of marked price = 120

Or marked price = $120 * \dfrac{100}{80}$

Or Markup = 50%

Correct answer choice (3)

Average

Averages

It is defined as ratio of sum of quantities to number of quantities

$$Average = \frac{x_1 + x_2 + \cdots + x_n}{n}$$

Problem Set

Question

The average monthly salary of 12 workers and 3 managers in a factory was Rs. 600. When one of the manager whose salary was Rs. 720, was replaced with a new manager, then the average salary of the team went down to 580. What is the salary of the new manager?

1. Rs 570
2. Rs 420
3. Rs 690
4. Rs 640
5. Rs 610

Correct Answer - **Rs 420**. Choice (2)

Explanatory Answer

The total salary amount = 15 * 600 = 9000

The salary of the exiting manager = 720.

Therefore, the salary of 12 workers and the remaining 2 managers = 9000 - 720 = 8280

When a new manager joins, the new average salary drops to Rs.580 for the total team of 15 of them.

The total salary for the 15 people i.e., 12 workers, 2 old managers and 1 new manager = 580 *15 = 8700

Therefore, the salary of the new manager is 9000 - 8700 = 300 less than that of the old manager who left the company, which is equal to 720 - 300 = 420.

An alternative method of doing the problem is as follows:

The average salary dropped by Rs.20 for 15 of them. Therefore, the overall salary has dropped by 15*20 = 300.

Therefore, the new manager's salary should be Rs.300 less than that of the old manager = 720 - 300 = 420.

Correct answer choice (2)

Question
The average wages of a worker during a fortnight comprising 15 consecutive working days was Rs.90 per day. During the first 7 days, his average wage was Rs.87 per day and the average wage during the last 7 days was Rs.92 per day. What was his wage on the 8th day?
    1. 83
    2. 92
    3. 90
    4. 97
    5. 96
Correct Answer - **97**. Choice (4)
Explanatory Answer
The total wage earned during the 15 days that the worker worked = 15 * 90 = Rs. 1350.

The total wage earned during the first 7 days = 7 * 87 = Rs. 609.
The total wage earned during the last 7 days = 7 * 92 = Rs. 644.

Total wage earned during the 15 days = wage during first 7 days + wage on 8th day + wage during the last 7 days.

ð   1350 = 609 + wage on 8th day + 644
ð   Wage on 8th day = 1350 - 609 - 644 = Rs.97.

Correct answer choice (4)

Question
The average of 5 quantities is 6. The average of 3 of them is 8.
What is the average of the remaining two numbers?
    1.   6.5
    2.   4
    3.   3
    4.   3.5
    5.   4.2
Correct Answer - 3 . Choice (3).
Explanatory Answer
The average of 5 quantities is 6.
Therefore, the sum of the 5 quantities is 5 * 6 = 30.

The average of three of these 5 quantities is 8.
Therefore, the sum of these three quantities = 3 * 8 = 24

The sum of the remaining two quantities = 30 - 24 = 6.

Average of these two quantities = $\frac{6}{2}$ = 3.
Note
From the answer choices, you can eliminate choices (1) and (2)
even without solving the question.

As the average of the 5 quantities is 6 and the average of three of
these five is 8, the average of the remaining two should be a value
that is less than 6. So choice (1) which is more than 6 can be
eliminated.

If there were a total of four quantities and the overall average was 6 and the average of 2 of the four were 8, then the average of the remaining two would have been 4. i.e., the simple average of 4 and 8 is 6. However, we have unequal number of quantities. Hence, choice (2) can be eliminated.

Correct answer choice (4)

Question
The average temperature on Wednesday, Thursday and Friday was 25^0. The average temperature on Thursday, Friday and Saturday was 24^0. If the temperature on Saturday was 27^0, what was the temperature on Wednesday?
1. 24^0
2. 21^0
3. 27^0
4. 30^0
Correct Answer is 30^0. Correct Choice is **(4)**

Explanatory Answer

Total temperature on Wednesday, Thursday and Friday was $25 * 3 = 75^0$

Total temperature on Thursday, Friday and Saturday was $24 * 3 = 72^0$

Hence, difference between the temperature on Wednesday and Saturday = 3^0

If Saturday temperature = 27^0, then Wednesday's temperature = $27 + 3 = 30^0$

Question

The average age of a group of 12 students is 20 years. If 4 more students join the group, the average age increases by 1 year. The average age of the new students is

1. 24
2. 26
3. 26
4. 22

Correct Answer is **24 years**. Correct Choice is **(1)**

Explanatory Answer

Total age of 12 students = 12 * 20 = 240 and the total age of 16 students = 21*16 = 336.

Let the average age of 4 new students be x.

Therefore total age of the new students = 4x.

Hence the total age of 16 students = 240 + 4x = 336 => x = 24.

Question

When a student weighing 45 kgs left a class, the average weight of the remaining 59 students increased by 200g. What is the average weight of the remaining 59 students?

1. 57
2. 56.8
3. 58.2
4. 52.2

Correct Answer is **57 kgs**. Choice **(1)** is right.

Explanatory Answer

Let the average weight of the 59 students be A.
Therefore, the total weight of the 59 of them will be 59A.

The questions states that when the weight of this student who left is added, the total weight of the class = 59A + 45
When this student is also included, the average weight decreases

26

by 0.2 kgs.

$$\frac{59A+45}{60} = A - 0.2$$

=> 59A + 45 = 60A - 12
=> 45 + 12 = 60A - 59A
=> A = 57.

Question

Three math classes: X, Y, and Z, take an algebra test.
The average score in class X is 83.
The average score in class Y is 76.
The average score in class Z is 85.
The average score of all students in classes X and Y together is 79.
The average score of all students in classes Y and Z together is 81.

What is the average for all the three classes?

1. 81
2. 81.5
3. 82
4. 84.5

Correct Answer is **81.5**. Choice **(2)** is right.

Explanatory Answer

Average score of class X is 83 and that of class Y is 76 and the combined average of X and Y is 79.

By rule of alligation ratio of students in X : Y is given by

$$X \qquad : \qquad Y$$

$$79$$

$$/ \qquad \backslash$$

$$83 \qquad\qquad 76$$

$$3 \qquad : \qquad 4$$

Similarly, average score of class Y is 76 and that of class Z is 85 and the combined average is 81.

By rule of alligation ratio of students in Y : Z is

Y   :   Z

81

/   \

76   85

4   :   5

∴ X : Y : Z = 3 : 4 : 5

∴ Total average for X, Y and Z = $\dfrac{3*83+4*76+5*85}{3+4+5}$

= $\dfrac{249+304+425}{12}$ = 81.5

Question

The average weight of a class of 24 students is 36 years. When the weight of the teacher is also included, the average weight increases by 1kg. What is the weight of the teacher?

1. 60 kgs
2. 61 kgs
3. 37 kgs
4. None of these

Correct Answer - **61 kgs**. Correct Choice is **(2)**

Explanatory Answer

The average weight of a class of 24 students = 36 kgs.

Therefore, the total weight of the class = 24 * 36 = 864 kgs

When the weight of the teacher is included, there are 25 individuals.

The average weight increases by 1kg. That is the new average weight = 37 kgs.

Therefore, the total weight of the 24 students plus the teacher = 25 * 37 = 925

Weight of the teacher = Weight of 24 students + teacher - weight of 24 students

= 925 - 864 = 61 kgs.

Question

The average of 5 quantities is 10 and the average of 3 of them is 9. What is the average of the remaining 2?

1. 11
2. 12
3. 11.5
4. 12.5

Correct Answer is **11.5**. Choice **(3)** is right.

Explanatory Answer

The average of 5 quantities is 10.
Therefore, the sum of all 5 quantities is 50.

The average of 3 of them is 9.
Therefore, the sum of the 3 quantities is 27.

Therefore, the sum of the remaining two quantities = 50 - 27 = 23.
Hence, the average of the 2 quantities = 23/2 = 11.5.

Question

The average age of a family of 5 members is 20 years. If the age of the youngest member be 10 years then what was the average age of the family at the time of the birth of the youngest member?

1. 13.5
2. 14
3. 15
4. 12.5

Correct Answer is **12.5**. Correct Choice is **(4)**

Explanatory Answer

At present the total age of the family = 5 * 20 = 100

The total age of the family at the time of the birth of the youngest member = [100-10-(10*4)] = 50

Therefore, average age of the family at the time of birth of the youngest member = 50/4 = 12.5.

Question

A student finds the average of 10 positive integers. Each integer contains two digits. By mistake, the boy interchanges the digits of one number say ba for ab. Due to this, the average becomes 1.8 less than the previous one. What was the difference of the two digits a and b?

1. 8
2. 6
3. 2
4. 4

Correct Answer - **2**. Choice **(3)** is right.

Explanatory   **Answer**

Let the original number be ab i.e., (10a + b). After interchanging the digits, the new number becomes ba i.e., (10b + a).

The question states that the average of 10 numbers has become 1.8 less than the original average. Therefore, the sum of the original 10 numbers will be 10*1.8 more than the sum of the 10 numbers with the digits interchanged.

i.e., 10a + b = 10b + a + 18, 9a - 9b = 18, a - b = 2.

Question

Average cost of 5 apples and 4 mangoes is Rs. 36. The average cost of 7 apples and 8 mangoes is Rs. 48. Find the total cost of 24 apples and 24 mangoes.

  1. 1044
  2. 2088
  3. 720
  4. 324

Correct Answer is **2088**. Choice **(2)** is right.

Explanatory Answer

Average cost of 5 apples and 4 mangoes = Rs. 36
Total cost = 36 * 9 = 324

Average cost of 7 apples and 8 mangoes = 48
Total cost = 48 * 15 = 720

Total cost of 12 apples and 12 mangoes = 324 + 720 = 1044
Therefore, cost of 24 apples and 24 mangoes = 1044 * 2 = 2088

Question

Average weight of 25 boys in a class is 48 kgs. The average weight of the class of 40 students is 45 kgs. What is the average weight of the 15 girls in the class?

  1. 44 kgs

  2. 42 kgs

  3. 40 kgs

  4. 39 kgs

  5. 42.5 kgs

Correct Answer - **40 kgs**. Choice (3)

Explanatory Answer

Total weight of boys in the class = 25 *48 = 1200 kgs

Total weight of all students in the class = 45 * 40 = 1800 kgs

Total weight of the girls in the class = 1800 - 1200 = 600 kgs

Average weight of girls = $\dfrac{600}{15}$ = 40 kgs
Correct answer choice (3)

Ramesh analysed the monthly salary figures of five vice presidents of his company. All the salary figures are in integer lakhs. The mean and the median salary figures are Rs. 5 lakhs, and the only mode is Rs. 8 lakhs. Which of the options below is the sum (in Rs. lakhs) of the highest and the lowest salaries?

 A. 9

 B. 10

 C. 11

 D. 12

 E. None of the above

Correct Answer : Rs. 9 lakhs. Choice (A)

Explanatory Answer

The mean salary of the five vice presidents is Rs.5 lakhs.
So, the sum of their salaries = 5 * 5 = 25 lakhs.

Let their salaries in ascending order be a, b, c, d and e.
So, a + b + c + d + e = 25.
The median salary is Rs.5 lakhs. So, c's salary is Rs.5 lakhs.

32

The only mode is Rs.8 lakhs.
So, Rs.8 lakhs salary is drawn by the maximum number of VPs.
C's salary is Rs.5 lakhs. So, d and e have to draw Rs. 8 lakhs each.

Therefore, $a + b + 5 + 8 + 8 = 21$ or
$a + b = 4$

Their salaries are in integer lakhs.
Therefore, a can draw Rs.1 lakh and b can draw Rs.3 lakhs or a and b can both draw Rs. 2 lakhs each.

However, there is only one mode. So, a and b cannot draw Rs.2 lakhs each.
So, a draws Rs.1 lakh (the least salary) and e draws Rs.8 lakhs (the highest salary).

So, $a + e = $ Rs.9 lakhs

Tips on Average

- If each number is increased or decreased by a certain quantity, then the average also increases or decreases by the same quantity.

- If each number is multiplied or divided by a certain quantity, then the average also gets multiplied or divided by the same quantity.

Ratio and Proportion

Ratio and Proportion

Ratio

Ratio is a relation between two quantities or numbers. A ratio of **a** and **b** is denoted by **a:b** and is read as: **a is to b**. In a ratio, the first part (a) is called Antecedent and second part (b) is called Consequent.

Proportion

Proportion is a statement that two ratios are equal. When two ratios are equal, the four terms involved, taken in order are called proportional, and they are said to be in proportion. a/b = c/d

Continued Proportion

Three quantities are said to be in continued proportion, if the ratio of the first to the second is same as the ratio of the second to the third. a/b = b/c; b is called mean proportion.

Compounded Ratio of two ratios a/b and c/d is ac/bd or ac : bd.

Invertendo: If a : b :: c : d then b : a :: d : c

Alternendo: If a : b :: c : d then a : c :: b : d

Componendo: If a : b :: c : d then (a +b) : b :: (c +d) : d

Dividendo: If a : b :: c : d then (a - b) : b :: (c - d) : d

Componendo and Dividendo: If a : b :: c : d then (a +b) : (a - b) :: (c +d) : (c - d)

I hope you enjoyed this. The idea is take each topic and practice as many problems as possible. The next edition we will showcase some more topics.